Expert's Guide to BYOD (Bring Your Own Device) in Training!

17 Ways to Create Learner Engagement and Increase Retention and Transfer using a single app!

By

Bob Pike, CPLP Fellow, CSP, CPAE

Published by CTT Press
1003 E. McNair Drive
Tempe, AZ 85283
612-747-5918
ISBN Paperback: 978-1-935291-02-2
ISBN eBook: 978-1-935291-05-3

TABLE OF CONTENTS

Introduction

Engagement is the word used today, but over the course of my career we called the same thing involvement, participation, and interaction among other things. What is clear though is that when people are really engaged in the learning process they learn faster, retain more, and apply more. And that is the true purpose of any training – to get results from what was learned – back on the job. And this makes training a process, not an event. It begins before training is ever delivered, in whatever form, and it continues until we see results in the workplace.

This book fits within a framework that I developed during the course of my career as a professional trainer that started back in 1969. The current iteration of that framework is called "Results-based Creative Learning Strategies".

Engagement is not something that can simply be tossed into a training program occasionally to wake people up or to refocus their attention. It is something that has its greatest effect when it is part of a systematic approach. Here are two fundamental principles that when understood and applied make engagement the powerful tool that it can and should be.

Principle 1 – my 90/20/8(4) rule. We know from research that adults can listen with understanding for up to 90 minutes. But they only listen with retention for 20 minutes. Because of the impact of media in the last 50 years (commercial television programs usually go no more than 8 minutes without a break and in the United States the average high school student has spent 19,000 hours watching television compared to 14,000 hours in class by the time they graduate) we need to engage participants in face to face training every 8 minutes. In virtual training engagement must occur every 4 minutes. So all content is chunked into a maximum of 20 minute chunks – and within those chunks engagement occurs at least every 8 minutes face to face, 4 minutes virtually.

Principle 2 – Successful engagement requires variety. A technique when used the first time is interesting, perhaps even exciting. But when used without interspersing other engagement techniques it rapidly becomes boring. During my career I've learned and adapted over 60 engagement techniques, but the 17 in this book are utilized most effectively when embedded within technology. This is because technology helps us to readily collect and store data, allows for easy capture and retrieval of ideas, allows key ideas to easily be revisited over time, and allows for capturing comments and voting.

Throughout my lifetime I have always been on the leading edge of technology. I was using a computer-based presentation system called "VideoShow™" along with software called "PictureIt™" in 1985.   When PCs became more broadly available I used

"Freelance™" and "Harvard Presentation Graphics". In the early 1990s I used wired audience response technology to get real time feedback during training, keynotes, and as a meeting facilitator. Then when wireless audience response systems became available I was an early adopter.

I met Stephen Halpern, who wrote "Tuning the Human Instrument" and who at that time had already spent 25 years researching how music impacts people. Based on what I learned from him I started working with my producer to create music to use during training programs that fit with his 25 years of research into how music affects people.

Fast forward to 2015. I learned of UMU and started using it in my conference keynotes and other training – both in the U.S. and internationally. Then I began building it into my training programs. The rest is history. It was a single piece of technology – a cloud-based app that allowed me to do, and do better, what I had been using 3 or 4 different pieces of software and related hardware to do. It also took advantage of what had now become ubiquitous – tablets and smartphones – so that I eliminated a large technology cost that I had previously incurred in order to create engagement. So let's dive in so you can get started!

CHAPTER 1

When can you use technology to increase engagement, retention, and transfer?

Since training is a process, not an event, then regardless of my delivery method I can divide the process into before, during, and after. I look at how I can engage both participants and managers before the training ever starts, during the training itself, and after the training is delivered to reinforce and support use of the new skills and knowledge on the job.

As you read the various chapters that drill down on engagement methods for a particular delivery method, always keep in mind that I want to connect and engage with both managers and participants before the training takes place, during the training, and after the training. Why? Because the engagement and connection before training sets the participant up for success both during the training and when they are back on the job looking to apply what they've learned in the real world.

CHAPTER 2

Types of Engagement – Variety is the Spice of Training!

In the appendix I've provided a chart that cross references the type of engagement with the delivery methods that it will work with. So mastering one type of engagement method will allow you to use it in a variety of situations! I've also provided a separate chart for development purposes showing what activities can be done on a mobile device or tablet and which can be done on a personal computer.

1. **Questions and Answers (QA).** This is probably the easiest of engagement methods and the one most often misused. Never ask, "Are there any questions?" Never ask for questions just before a break, lunch, or at the end of the day. At least not if you actually want a question! Ideally, I want to give people reflection time – a minute or two to come up with a question. Often, I have small groups brainstorm

questions, then the groups ask a question. When using UMU I can have participants use their smartphones or tablets and log into my course. They can then enter their question and when they press submit they see all the questions being asked and give a "thumbs up" to the ones they like. Then I can see all the questions in rank order of "likes". Typically, I'll answer the top 3-5 questions depending on time. Later I can either go back during class and answer additional questions - or record answers after class and push the answers out to the participants. Pretty cool, isn't it?!

2. **Polling/Survey (PO).** Polling allows me to have people vote on content I provide while a survey allows me to gather information from them. For example, in my class on "Building a Multimillion Dollar Training/Consulting Business" I use both polls and surveys. Before the class I send a link to a survey in UMU. Participants answer questions about where they are in their business (from working alone to multiple employees), look at a list of the ten topics in the class vote for the top 3, and then answer an open-ended question – why they are choosing to participate in this program. This allows me, before class begins to have some understanding of the participants and to look at how I might shift my time structure. Here are what some of the results look like on UMU:

81	90	90.0%
Responded	Participated	Response Rate

Q1 Where are you in your business? (Choose all that apply) (Multiple Select)

A. Haven't started
12.3% 10

B. Just me
59.3% 48

C. Me plus support staff
25.9% 21

D. Passive Income
14.8% 12

E. Have employee consultants
8.6% 7

F. Have freelancers
18.5% 15

Q2 Choose your top 3 topics for this session (Multiple Select)

A. Acquiring talent to deliver my content
14.8% 12

B. Transferring Expertise
30.9% 25

C. Powerful Proposals
70.4% 57

D. Passive Income
49.4% 40

E. Content Creation
46.9% 38

F. Fast Track Publishing
33.3% 27

G. Staying on top of your game.
46.9% 38

5. To learn how I can create content with purposeful transference of expertise, establish passive income from such effort, and create powerful proposals to lock up deals and meaningful relationships with prospective clients

6. I've been a Bob Pike fan for many years. I always am energized after one of his presentations. I'm looking for more great ideas to continue to improve my design and delivery of workshops.

7. I am now preparing my seminars based upon my eBook Blockbuster Branding For "The Little Guy". My focus is on launching and growing my business.

8. Open to learn to enhance my speaking business...and I wanted to visit my NSA-Arizona chapter (live in Canada - so don't get down often)

9. I want to learn strategies you used Bob in your world that I can transfer to use in mine. That's the short answer. (Dan Shinder)

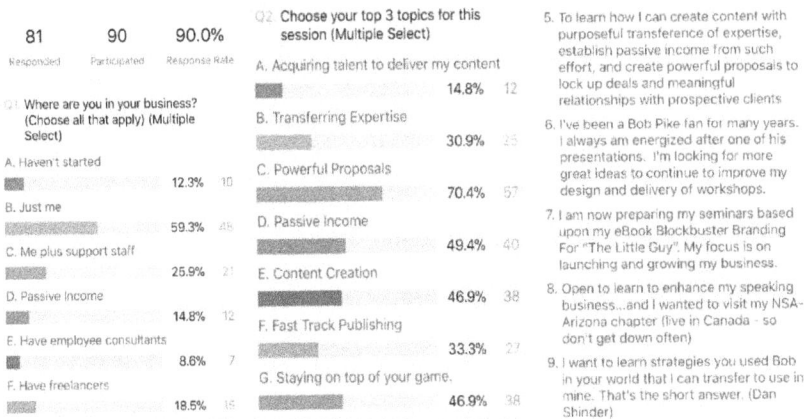

3. **Evaluation (EV)**– When I say evaluation I'm not only looking at Kirkpatrick's Level 1: Reaction (Did they like it?) I'm also looking at using technology to measure levels 2 Learning (Did the learn it?), Level 3 Behavior (Did they use it?), and Level 4 Results (Did it make a difference?).

During class I'll use UMU to get intermediate feedback using a very simple open ended Start, Stop, Continue. What's something that we should start doing that would make the class more valuable? What's something that we should stop doing – is it getting in the way of learning? What's something we should continue doing – is it working well? This allows me to adjust the class based on feedback in real time from participants to maximize learning and transfer.

Periodically I'll use a quiz to evaluate learning. After the class is over I'll use UMU to reach out to both participants and managers about what people are doing differently (and what the impact (results) are (Levels 3 and 4).

4. **Discussion Board (DB)**– for me the ideal discussion in face to face learning takes place when I create groups of 5-7, have them choose a group leader, provide them with discussion questions, give them a time limit and let them go. But I can also use UMU to create a discussion by posting a discussion question and then allowing participants to record their responses (either with their comment identified or anonymously), they then reply to other comments that have been made, and can vote for the comments they see so that the final result is on a rank ordered screen. I can use this for either synchronous delivery (when everyone is participating at the same time) or asynchronously (when people are participating at different times).

5. **Raffle (RA)**– I always have participants turn in evaluation forms early. Why? Because the worst way to end any training program delivered by any method is to ask people to fill out an evaluation form. 90% of the classes I've observed for clients ended with an evaluation form. And what participants want to do is leave. So I begin the class by telling people what they will be providing feedback on during the course of the class. It usually involves evaluating four things: the instructor, the content, the materials and environment, and the participants themselves (did they contribute to the learning of others around them, etc.) I use a raffle near the end of class to incentivize everyone to have their evaluations ready to turn in.

6. **Flip Chart (FC)** If you've ever been in one of my face to face classes you know that my preferred visual aid is the flip chart. Why? Because it can be prepared by either the instructor or the participants. It can be prepared before the class – or created in real time during the class. But the thing that elevates the flip chart above other visual aids is that once created it can hang on the wall for the duration of the class and serve as a reminder of key learning points. Once a PowerPoint is used it's gone, the same holds true for video. But the flip chart endures. In all my classes there will be participants taking photos of various flip charts. I can do the same thing with UMU. I can take pictures of the flip charts, turn them into slides and project them to the screen. At the same time they are saved in UMU so I can make them available after the class is over.

7. **Attendance (AT)**– For each session of a class I want participants to sign in. Using UMU I can have them enter their name, gender, email, and cell phone number. In subsequent class sessions I have them enter their name and a question they want to ask about previous content or the most important thing they learned in the previous session. In either case they can vote on the content the class provides and I can use the results as a great session opener.

8. **Quiz (QU)**– A great way to periodically reengage learners and to check for knowledge is to give a quiz. UMU allows me to create either single answer or multiple answer questions. I can also assign various point values to questions.

9. **Game Break (GA)**– UMU has 4 quick built-in logic/reasoning games that I can use to refocus attention and rebuild energy during a class.

10. **Video (VI)** – I can either upload prerecorded videos or record videos "on the fly" within the app. This allows me to have participants watch video before, during, or after a training event. I can also create short "how to" videos. Some participants learn better by being able to watch something several times rather than simply reading or listening about a process.

11. **Slides with Audio (SA)** I can narrate PowerPoint slides and upload them to my UMU app. Or, I can take an assortment of flipcharts and photos that I've captured with UMU and create a narration around them.

12. **Readings (RD)** There are times when I want participants to read something, reflect on it, and make comments – or ask a question. Again, this may be before, during, or after a class. UMU allows me to do all of these things.

13. **Video Coaching (VC)** One of my former trainers, Sue Enz, once said, "Just because I said it, doesn't mean that you learned it." This is so true. And equally true is "Just because you've seen it, doesn't mean you can do it." I learned this the hard way when it came time to learn to drive a car. I had watched my parents for years, so I was sure that I could do it. And then I got behind the wheel of our 1954 Hudson with a 3-speed stick shift transmission. I started the car and tried to put it into reverse without depressing the clutch. There was a loud metallic grinding sound. I depressed the clutch, put it into reverse, and then lifted the clutch without adding any

gas. Another loud grinding sound, the car lurched backward, and the engine died. My Dad said, "One more move like that and your driving is over." I was painfully aware that though I thought I knew how to drive, I didn't. Video coaching is what I call it when I add short video clips that walk-through things that participants have learned in class – so that they can, as often as necessary, remind themselves of what good looks like.

14. **Action Planning (AP)** In most training programs we don't give participants enough reflection time – time to think about what they are learning and how they'll use it. If people don't plan for application during training, chances are that they won't once the training is over. With the UMU app I can give them reflection time and have them post their action ideas. Once they submit, they see everyone's action ideas and "like" any that they choose to. This allows them to build and refine their own action plans.

15. **Leaderboard (LB)** With my UMU app I can create a series of questions allowing participants to score points throughout a learning event. They can also score points for completing activities, commenting, asking questions, etc.

16. **Infographic (INF)** I can combine 3-5 photos or flipcharts along with additional captions to create a single infographic that can convey key information.

17. **Notification (NO)** I can use my UMU app to send out a variety of notifications – new material available, projects or activities that are due, a discussion that participants need to engage in, etc.

There are, of course, other ways that I could get these things done and other tools or technologies that I could use – but the UMU app lets me do all of these with one simple, cloud-based app – and that's exciting!

CHAPTER 3

Increasing Engagement Through Prework

Too many trainers teach classes where participants simply "show up". If I had a dollar for every person attending a class whose answer was "I was sent." in answer to the question, "Why are you here?" I'd be a very rich man! I want to engage both participants and the managers who are sending them because this sets up their training for success.

I have a variety of ways that I can get them engaged before the class even starts. Here are four to get started:

1. I can send them a questionnaire to be completed before class. This allows me to find out more about them, why they are attending, how their manager has prepared them for the training, their familiarity with the course content, their learning objectives for the course, etc.

2. I can send them a survey. In a recent class I asked each participant to choose their top 3 (of 9) topics for the upcoming class. This enabled me to get a sense of what was important for each participant and perhaps to adjust the amount of time spent on each topic.

3. I can ask them a question or series of questions. I can ask them for 3 questions they want answered by the time the class is completed, or the most important skills they want to master, or how they plan to use what they've learned back on the job. You get the idea.

4. I can send them an article to read. UMU makes it easy to upload several articles that I can push to my participants all at once or over time. There are many things that I might want them to read before the class starts. It might be a short piece that positions the class. It might be an article by a subject matter expert. It might be a letter from a high-ranking executive within the company outlining the importance of the class. My goal is to engage the learner and to create excitement for the class.

5. I can also send them a pretest. This allows me to set a baseline before the class begins. Some might argue that sending them a pretest might allow participants to "Cheat". However, if this causes them to learn then perhaps that's cheating that should be encouraged.

CHAPTER 4

Instructor-led engagement (IL)

I can use UMU in at least 10 different ways to put variety and engagement in my instructor led programs.

1. Question and Answer. At least 3 to 4 times per day I want to allow my participants reflection time to think of questions they want to ask. They then go to the UMU app and enter their questions. Once they've submitted their question they will see the entire groups questions on their smart phone. They are then able to "like" the other questions and UMU re-orders the questions based on the number of likes. I can then choose to answer questions for a period of time-say 12 minutes. At the end of 12 minutes I may have answered 3-5 questions. Later I can go into UMU and dictate audio answers to as many of the other questions as I want. I can then make those answers available to the participants.

2. Polling/Survey. I usually start the class with the survey. I might ask them about their level of experience or where they are from - something that helps both me and their classmates learn more about them. Throughout the class I use short surveys to gain insight into what they're learning, the pace of the class, etc.

3. Evaluation - I'm not going to wait until the end of the class to get feedback. I might use the start, stop, continue method for feedback. Many times, at the beginning of course-especially longer ones-I let participants create five guidelines they want me to follow to maximize the value of the course and five guidelines they're willing to follow to maximize the value of the course. Then periodically I'll use UMU to have them evaluate how well I'm doing in living up to the guidelines as well as how well the participants are doing in living up to the guidelines.

4. Raffles. Periodically I use the raffle activity to randomly select several members to win a small prize. Typically, this is after a break – and winners must be present!

5. Flip charts. I use a lot of flip charts in my instructor led training. And many of the participants take photos of them for later reference. What UMU allows me to do is to take photographs of those flip charts and build them into my UMU course so that all participants will have access to them after the course is over.

6. Attendance. UMU allows me to have participants login so that if I want to, I can maintain a record of attendance. This can be downloaded in CSV format so it can then be uploaded to a learning management system.

7. Quiz. Throughout the class I will use a quick quiz to test for knowledge and understanding. UMU allows me to ask single answer, multiple choice, and short answer questions.

8. Games. UMU has 4 quick interactive games built in that I can use to get participant energy back up and to quickly reengage learners after breaks.

9. Notifications. Especially in classes that are conducted over time UMU gives me the ability to send out messages to participants. This includes reminders of the next class, items to bring to class, assignments to be completed between classes, etc.

10. Generally, in using all of these activities I move between PowerPoint and UMU. I cue my audience by using slides like these::

Take the survey now!

**This session uses UMU
for audience interaction.**

Go to UMU.COM and enter PIN:

422

Ask your question now.

This session uses UMU for audience interaction.

Refresh your screen

OR

Go to UMU.COM
Enter PIN **422**

CHAPTER 5

Engagement in the Flipped Classroom

What is a flipped classroom? Basically, it is a blended learning approach that asks the question: What can participants learn on their own before actually meeting face-to-face in an instructor led classroom? So, a large chunk of knowledge acquisition is "flipped" from the face to face classroom to participants learning on their own before the class takes place.

It is designed to allow the face-to-face time to maximize skill practice and application of what is learned rather than taking up face-to-face time with the instructor delivering content that can be acquired by participants on their own, before class. Building my content for flipped learning in my UMU app allows for easy delivery to all participants as well as tools to insure that the content is absorbed, and the face to face time has the greatest effectiveness.

Almost all the features of UMU can be put to work when using a flipped classroom.

1. Questions and answers. Participants can submit questions as they work through material on their own. The instructor can dictate answers and push them out to the participants.

2. Polling/Surveys. I can periodically check on the participants progression with the content and the degree to which they are use a utilizing the content on the job.

3. Evaluation. I can use my stop/start/continue process to see how both the content and process are working for the participants.

4. Discussion Board. I post discussion questions for all participants to engage in for a period of time. All participants can see the comments and are able to like those they find useful.

5. Quiz. I periodically post quizzes for the various content segments to test for knowledge.

6. Videos. I utilize short videos to provide both knowledge into demonstrate skill that participants need to master.

7. I upload slides with an audio commentary to create short content chunks for participants.

8. Readings. I post short articles as another way to deliver content. I can also post files to download containing projects to work on and submit.

9. Video coaching. I create short videos demonstrating skills that participants can watch and re-watch to aid them in their own skills mastery.

10. Leaderboard. Participants score points for completing various assignments and can see their progress measured against other participants.

11. Infographics. Posting infographics gives me another way to deliver content as opposed to readings or videos and in some cases serve as excellent "cheat sheets" or job aids.

12. Notifications. I use my UMU app to inform participants that new material is available, projects and activities are due, and there are discussions to be engaged in, etc.

CHAPTER 6

Using Technology for Coaching

There are two places my UMU app can help me best when I'm coaching: Videos and Notifications.

1. Videos. I can break skills down into short video segments so that those I coach can view them as many times as needed to help master those skills.

2. Notifications. I can use UMU as my way to connect with someone I coach. I can also provide new material for them to download or complete. This allows me to keep my communications organized and provides a tracking mechanism.

Chapter 7 Technology in the Learning Lab (LL)

What is a learning lab? It is simply a room where participants are gathered in one place yet work independently to accomplish tasks. Here the 6 ways UMU can help me when I use this delivery method:

1. Notifications. I can notify all participants at the same time that a task needs to be done – especially the UMU activities that follow here.
2. Quiz. I can periodically check the progress of everyone in the lab with a quick quiz.
3. Video. All learning lab participants can view videos that help them with their learning lab goals as often as necessary.
4. Slides with Audio. I can narrate PowerPoint slides and upload them to my UMU app for the use of learning lab participants. Or I can take an assortment of flipcharts

and photos that I've created, load them onto UMU and create a narration around them.

5. Readings. I can post articles on the UMU app for learning lab participants to read. I can also provide "cheat sheets" or job aids for them to download.

6. Discussion Board (DB)– for me, the ideal discussion in face-to-face learning takes place when I create groups of 5-7. They choose a group leader. I provide them with discussion questions, give them a time limit and let them go. However, the learning lab is designed for people to learn on their own even though they are in the same place. However, I can still use UMU to create a discussion by posting a discussion question and sending out a notification. This allows participants to record their responses (either with their comment identified or anonymously), they then reply to other comments that have been made and can vote for (like) the comments they see so that the result is on a rank ordered screen.

CHAPTER 8

Self-Paced Learning (SP)

1. Quiz (QU) A great way to periodically reengage learners and to check for knowledge is to give a quiz. UMU allows me to create either single answer or multiple answer questions. I can also assign various point values to questions.

2. Upload prerecorded videos or record videos "on the fly" within the app. Either way participants can watch the video before, during, or after a training event. A short "how to" video can be created. Some participants learn better by being able to watch something several times rather than simply reading or listening.

3. Slides with Audio (SA) I can narrate PowerPoint slides and upload them to my UMU app. Or I can take an assortment of flipcharts and photos that I've created, load them onto UMU and create a narration around them.

4. Readings (RD) There are times when I want participants to read something, reflect on it, and comment – or ask a question.

5. Discussion Board (DB) I can post questions, etc. that I want participants to think about and comment on – UMU allows me to capture the comments, comment back – and have participants like comments so we can see what consensus building there may be.

6. Infographic (INF) UMU allows me to combine 3-5 photos or flipcharts along with additional captions to create a single infographic that can convey key information.

7. Notification (NO) UMU can send out a variety of notifications – new material available, projects or activities that are due, a discussion that participants need to engage in, etc.

CHAPTER 9

Virtual Learning (VL) (Webinars)

While many webinar platforms allow me to use the tools I've listed here, I love the unique way that UMU allows me to use them.

1. Questions and Answers. People can not only post their questions – they can vote for the questions other attendees have submitted. During the webinar I answer the 3-5 questions that get the most votes. After the webinar, however, I can use the app to record answers to as many of the other questions as I like. Bonus tip – do a Q&A about two thirds of the way through an hour-long webinar. If you wait until the end people leave the webinar and can miss real value.

2. Polling/Surveys Polling allows me to have people vote on content I provide while a survey allows me to gather information from them.

3. Quiz. Since webinars are generally shorter in time, I use the quiz primarily as a change of pace and as an energizer. It will be short – probably 3 questions at most.

4. Evaluation. Since webinars tend to be shorter I keep the evaluation short as well. How useful was the content? How knowledgeable was the instructor? Would you recommend this session to others?

5. Action Planning. Just after Q&A (two thirds of the way through an hour webinar,) I allow reflection time and then use the UMU app to allow people to list one or two action ideas from the session. Once they've submitted their action ideas they'll see all the others being posted and they "like" the ones they find useful. At the end, all ideas posted will be seen rank ordered. A great way to help stimulate thinking – and action!

CHAPTER 10

Using UMU for Performance Support

What can I do to support performance after skills and knowledge have been built? UMU provides several tools I can use for performance support.

1. Video (VI) – I can either upload prerecorded videos or record videos "on the fly" within the app. This allows participants to watch the video before, during, or after a training event. A short "how to" video can be created. Some participants learn better by watching something several times rather than simply reading or hearing about a process.

2. 2. Slides with Audio (SA) I can narrate PowerPoint slides and upload them to my UMU app. Or, I can take an assortment of flipcharts and photos that I've created, load them onto UMU and create a narration around them.

3. 3. Readings (RD) There are times when I want participants to read something, reflect on it, and comment on it – or ask a

question. At other times I want additional print materials – articles, diagrams, flow charts, etc. available after class. The UMU app does this for me.

4. 4. Discussion Board (DB) I can post questions, etc. that I want participants to think about and comment on – UMU allows me to capture the comments, comment back – and have participants like comments so we can see what consensus building there may be.

5. 5. Infographic (INF)I can combine 3-5 photos or flipcharts along with additional captions to create a single infographic that can convey key information.

6. 6. Notification (NO) I can use my UMU app to send out a variety of notifications – new material available, projects or activities that are due, a participant discussion, etc.

CHAPTER 11

Using UMU to Create Microlearning

Microlearning deals with relatively small learning units and short-term learning activities. Generally, the term "microlearning" refers to micro-perspectives in the context of learning, education and training. More frequently, the term is used in the domain of e-learning and related fields. Whatever learning is developed can be accessed not only from a computer, but also from a smartphone or tablet.

UMU provides me with the perfect platform for creating and delivering microlearning by:

1. Using Notifications so people regularly interact with content
2. I can make the content available over time, so they can't complete it all at once. If you learn it all at once, you forget it all at once.

3. Using a Leaderboard. This plays on competitive spirit and lets people know when they are falling behind in comparison to peers.

4. Videos that can be either upload or created from within the UMU app. These can be short demonstrations of skills or include other content.

5. Slides with Audio. I can create these with UMU – or create slides and record a voiceover in PowerPoint and the upload the results into UMU.

6. Quizzes – I can create multiple choice, choose all that apply, or a short answer. The point values can vary, and the results are shown on the Leaderboard.

7. Readings/files for download.

8. Discussion Board (DB) I can post questions, etc. that I want participants to think about and comment on. UMU allows me to capture the comments, comment back – and have participants "like" comments so we can see what consensus there may is.

CHAPTER 12

Capturing and Delivering Subject Matter Expert Knowledge (SME)

This approach is virtually identical to creating microlearning. The difference is where the content is coming from. You may want to capture knowledge from high performing employees, especially those nearing retirement, so that the knowledge and skills can be transferred to others. You can also use it for peer-to-peer learning – especially in sales and customer service to identify, capture and disseminate best practices. UMU allows you to bring internal experts together as content creator that you manage, edit, distribute, track, and retain with UMU.

The easiest way to capture expertise is with video. Interview the SME. After, you can edit the video into smaller slices and use that as part of a microlearning course capturing the SMEs expertise.

I can also videotape the SME performing a skill and providing a running explanation of what he/she is doing. I can then take the content I've captured and create articles, cheat sheets, infographics, etc. When I build in discussion boards, quizzes, and a leaderboard I've now created a way to better transfer SME knowledge to another person or to a group.

CHAPTER 13

Making Meetings Work with Technology (MMW)

1. Survey for Agenda Setting. UMU is a great way to make meetings work better. I start by creating an open-ended survey to collect agenda items in advance. I can also let people vote in advance to prioritize agenda items – and perhaps to scale back the agenda.

2. Q & A for Decision Making. I can use the Q & A activity to allow people to vote on items. The vote can be either public or anonymous.

3. Q & A for Brainstorming. I can use the Q & A activity to allow participants to brainstorm, comment, and vote on ideas – all anonymously. This helps people to contribute even with superiors present when they might not otherwise.

4. Polling/Survey During the meeting I can use a survey to collect and visualize opinions. Participants can then vote and UMU allows me to display the prioritized results.

When the meeting is over I can add notes and an executive summary and create a polished pdf report.

CHAPTER 14

Using Technology for Quick Updates (QU)

UMU is ideal for creating quick updates about either software or other products. Business is constantly changing and innovating. It's difficult to keep everyone on the same page. UMU allows me to quickly create content I can track without sacrificing knowledge about employee participation and performance. UMU helps employees keep pace with the changes in the organization. No one is left behind!

Use these simple steps:

1. Create a microlearning course. Combine short chunks of information into a complete course.
2. Add activities:

    a. Track attendance
    b. show video clips

c. Survey / Polling - Create single-question polls or multi-question surveys.

d. Q&A / Word Cloud - Let your audience suggest questions and vote to crowdsource the best ones. Answer them periodically and push the answers out to everyone signed up for the update course you've created.

e. Quiz - Give participants a graded quiz. Advanced settings allow me to adjust point values and set time limits.

f. Discussion/Word Cloud - Create a response feed that gets everyone on the same page.

g. Flipchart Slides - Instantly stream photos to the screen and store them in your account. A great way to save flip charts!

CHAPTER 15

UMU – the All-in-One Technology Solution

I've been using UMU now since the middle of 2015. I used it first at a keynote in Malaysia. It added a WOW factor to my presentation and allowed me to stay connected with the entire audience afterwards.

Next, I used it in a training program and keynote at China's Training Magazine Conference in Shanghai. Again it added a WOW factor – and after the keynote the UMU staff helped translate the questions and I recorded answers which they then translated into Chinese as a way to further support the 900 trainers that were in the audience. (As an aside it was here that I asked what UMU meant and was told that it didn't stand for anything. Always wearing my marketing/branding hat I said, "It does stand for something – it stands for U, Me, and Us because of all the ways it helps you, and me, and all of us to connect with each other!") And that explanation stuck!

UMU has become my technology of choice. I have many technology tools in my learning and development arsenal, but this is my learning technology swiss army knife (it has an amazing number of tools that can be used in an amazing number of situations)! It does so many more things than any single piece of technology – and it does them well.

UMU as an app is organized very simply: Content, Courses, and Learning Programs. Course are groups of content elements that the instructor puts together. Learning programs are courses that I group together. Once I create content it can be used in multiple courses. When I create a course it can be grouped into multiple learning programs. What a fantastic way to leverage the time you invest in developing your content when you use UMU!

Content is what you contribute to the UMU app. Content elements are the frames UMU provides to contain your content. There are two basic categories of content elements: Microlearnings and Activities.

Microlearning elements are: Audio Slides, Video, Article, Infographic, File (Readings), and Live Broadcast.

Activities include: Attendance, Q&A, Discussion, Flipchart Slides, Raffle Drawing, and Game Break.

The exciting thing is that once I create any content element I can save it as a template for easy re-use – either as is or with editing for a future course.

The key to using UMU is progress not perfection. Try something, test it out, try something else, add them together – and keep on going. I love UMU's collaboration feature. I can allow multiple people access to content I'm creating so that they can contribute to the creative process.

I've provided two helpful charts in the appendix. One shows what features of UMU can currently be accessed from a computer and which can be accessed from a mobile device or tablet. The other is a matrix showing the UMU activities across the top and where they can be used down the left-hand column.

So, let me challenge you not only to build the activities that I've outlined in this book into your training, but also use the UMU app to do it. I believe you'll be as excited about the results as I am.

UMU Features: App vs. Desktop

UMU is the only all-in-one microlearning platform that supports content creation from both desktop and app platforms. Just as you may prefer your phone over your computer in certain scenarios, we've done our best to optimize our content builders to the device that suits it best. In the future, each of these features will be completely synchronous.

		Desktop Platform	App
Create	Audio Slides, Infographic	X	√
	Live Broadcast	X	√
	Video	√	√
	Article	√	X
	Survey, Quiz, Attendance, Flipchart Slides, Q&A, Discussion, Game Break	√	√
	Raffle Drawing	√	X
Learning Group	Text and pictures	√	√
	Audio and video messages	X	√
	Block messages	X	√
	Adjust advanced settings	X	√
	Send content cards	X	√
Course Management	Collaborate	√	√
	Create & Manage Templates	√	√
	Edit Course Info and Settings	√	√
Data & Reporting	Export Activity Data to Excel, PDF, and Word	√	X
	Export Videos & Audio Slides	√	√
Learner Management	Manage Registrations and Monitor Responses	√	√
	Grade Open-ended Quiz Questions	√	Coming Soon
	View Learner Feedback, Participation, Likes, and Ratings	√	√
Inbox	View Course Updates, Invitations, and Replies	X	√

Methods

Methods (types of engagement) from Chapter 2

WHERE	QA	PO	EV	DB	RA	FC	AT	QU	GA	VI	SA	RD	VC	AP	LB	INF	NO
PW								X		X	X			X			X
IL	X	X	X	X	X	X	X	X	X					X			X
FC	X	X	X	X	X	X	X	X	X	X	X	X		X	X	X	X
C								X									X
LL		X					X	X		X	X	X		X			X
SP								X		X	X	X					X
VL	X	X	X	X	X			X									X
PS										X	X	X				X	X
ML										X	X	X	X	X			
SME								X		X	X	X					
MMW	X	X	X	X	X	X								X			X
QU								X		X	X	X					X

About the author

Bob Pike CPLP Fellow, CSP, CPAE-Speakers Hall of Fame has long been known globally as the "trainer's trainer". More than 150,000 trainers on five continents have graduated from his train the trainer programs. He has designed more the 600 training programs of one day or longer for 80% of the Fortune 1000. He is the author of more than 30 books in the learning and development field including *Dealing with Difficult Participants, 50 Creative Openers, One on One Training, The Fun Minute Manager,* and the all time best-selling train the trainer book ever published *Creative Training Techniques Handbook (now titled Master Trainers Handbook)* which has sold over 330,000 copies in 4 editions. He has delivered keynotes, training, and consulting in more than 25 countries.

He is the founder/editor of *Training and Performance Forum* a 12-16-page digital newsletter of training tips, techniques, and tactics that he has provided to trainers globally since 1989.

He can be reached at Bob@CTTNewsletters.com

www.ingramcontent.com/pod-product-compliance
Lightning Source LLC
Chambersburg PA
CBHW071326200326
41520CB00013B/2884